DID YOU KNOW?
ANIMAL ADAPTATIONS

20 THINGS YOU DIDN'T KNOW ABOUT

SEA ANIMAL ADAPTATIONS

SLOANE HUGHES

PowerKiDS press
New York

Published in 2023 by The Rosen Publishing Group, Inc.
29 East 21st Street, New York, NY 10010

Portions of this work were originally authored by Tayler Cole and published as *20 Fun Facts About Marine Animal Adaptations*. All new material in this edition was authored by Sloane Hughes.

Editor: Amanda Vink
Book Design: Tanya Dellaccio

Photo Credits: Cover Jung Hsuan/Shutterstock.com; p. 5 (dolphin) muratart/Shutterstock.com; p. 5 (starfish) Vojce/Shutterstock.com; p. 5 (octupus) Bildagentur Zoonar GmbH/Shutterstock.com; p. 6 KT photo/Shutterstock.com; p. 7 HikoPhotography/Shutterstock.com; p. 8 https://upload.wikimedia.org/wikipedia/commons/5/5b/Vampyroteuthis_photo.png; p. 9 Doug Perrine/Alamy Images; p. 10 Blue Planet Archive/Alamy Images; p. 11 (salps) Damsea/Shutterstock.com; p. 11 (squid) RobJ808/Shutterstock.com; p. 11 (jellyfish) Vojce/Shutterstock.com; p. 12 Sergey Uryadnikov/Shutterstock.com; p. 13 Christian Musat/Shutterstock.com; p. 14 valda butterworth/Shutterstock.com; p. 15 Anne Webber/Shutterstock.com; p. 16 Ethan Daniels/Shutterstock.com; p. 17 Rebecca Schreiner/Shutterstock.com; p. 18 aquapix/Shutterstock.com; p. 19 Wiskerke/Alamy Images; p. 20 Florida Chuck/Shutterstock.com; p. 21 Agung bayu/Shutterstock.com; p. 22 Steven L. Gordon/Shutterstock.com; p. 23 Sahara Frost/Shutterstock.com; p. 24 walkdragon/Shutterstock.com; p. 25 Gerald Robert Fischer/Shutterstock.com; p. 26 Barbara Ash/Shutterstock.com; p. 27 (stargazer) Tatiana Kilimnik/Shutterstock.com; p. 27 (cuttlefish) Richard Whitcombe/Shutterstock.com; p. 27 (crab) Daria Artemova/Shutterstock.com; p. 29 Irina Wilhauk/Shutterstock.com.

Library of Congress Cataloging-in-Publication Data
Names: Hughes, Sloane, author.
Title: 20 things you didn't know about sea animal adaptations / Sloane Hughes.
Other titles: Twenty things you didn't know about sea animal adaptations
Description: New York : PowerKids Press, [2023] | Series: Did you know? animal adaptations | Includes index.
Identifiers: LCCN 2021062469 (print) | LCCN 2021062470 (ebook) | ISBN 9781538386910 (library binding) | ISBN 9781538386897 (paperback) | ISBN 9781538386903 | ISBN 9781538386927 (ebook)
Subjects: LCSH: Marine animals–Adaptation–Juvenile literature.
Classification: LCC QL122.2 .H84 2023 (print) | LCC QL122.2 (ebook) | DDC 591.77–dc23/eng/20220106
LC record available at https://lccn.loc.gov/2021062469
LC ebook record available at https://lccn.loc.gov/2021062470

Manufactured in the United States of America

CPSIA Compliance Information: Batch #CSPK23. For Further Information contact Rosen Publishing, New York, New York at 1-800-237-9932.

CONTENTS

WATER WORLD

Our planet is largely covered with water. The oceans are huge bodies of water with many diverse **environments**. Over thousands of years, marine animals' bodies and behaviors have changed to help them better survive. These changes are called adaptations.

Scientists have recently discovered some kinds of marine animals. However, we've known about others for hundreds of years. Let's explore some of the amazing features that allow marine animals to spend their lives underwater!

OCTOPUS

Humans have discovered at least 226,000 marine species so far!

SEA STAR

DOLPHIN

MARINE BODIES

DID YOU KNOW?

A puffer fish has a special trick—it can turn its body into a balloon!

Puffer fish swim very slowly, which makes them seem like an easy meal to some predators. When **threatened**, a puffer fish will suddenly take in large amounts of water or air through its mouth. It quickly pumps the water (or air) into its stomach so it becomes a large ball that predators can't fit in their mouth.

Some puffer fish have spikes on their body. That helps protect them too!

DID YOU KNOW?

Puffer fish can be very **toxic**—but one kind of predator eats them anyway.

Most puffer fish produce a toxic **substance** that can make them deadly. In some places around the world, however, humans call puffer fish "fugu" and consider them a special treat. Trained chefs prepare it. However, people can be poisoned if they eat the wrong part.

GLOWING CREATURES

DID YOU KNOW?

The vampire squid uses bioluminescence to confuse predators.

The vampire squid doesn't shoot black ink like some squids and octopuses. Instead, it shoots a colorless substance that has bits of light-producing material. Because it's so dark in the deep ocean, the light confuses predators. When an animal creates visible light, it's called bioluminescence.

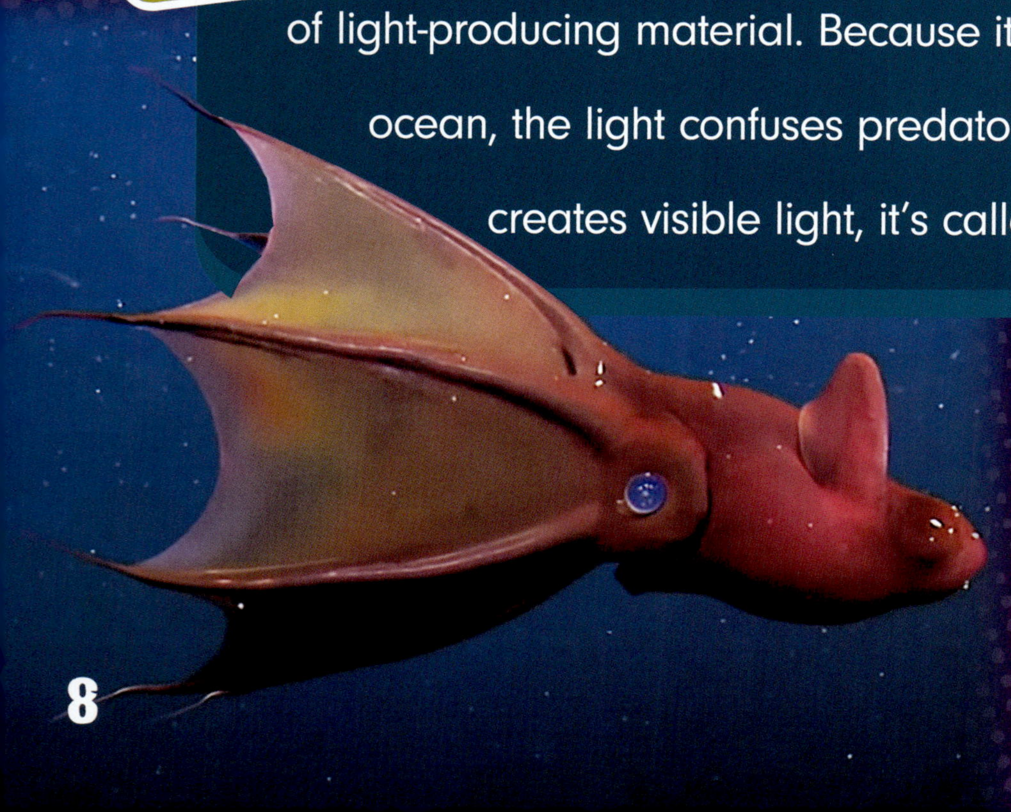

Male anglerfish attach themselves to a female's body and get all the **nutrients** they need from her. In time, their body becomes fused to the female's body.

DID YOU KNOW?

The female anglerfish uses bioluminescence to catch prey!

Light can't make it down to the bottom of the ocean, and that's where anglerfish live. Female anglerfish have developed their own "light bulb" that hangs above their mouth! Other fish come to check out this light, and that's when the anglerfish gobbles them up!

DID YOU KNOW?

Cookiecutter sharks have glowing bellies.

The entire underbody of this shark is covered in tiny light-producing organs that glow green. Scientists believe that this light helps to draw in prey. Cookiecutter sharks feed on small fish and larger marine animals.

Cookiecutter sharks take even, round bites out of larger animals—and that's how they got their name.

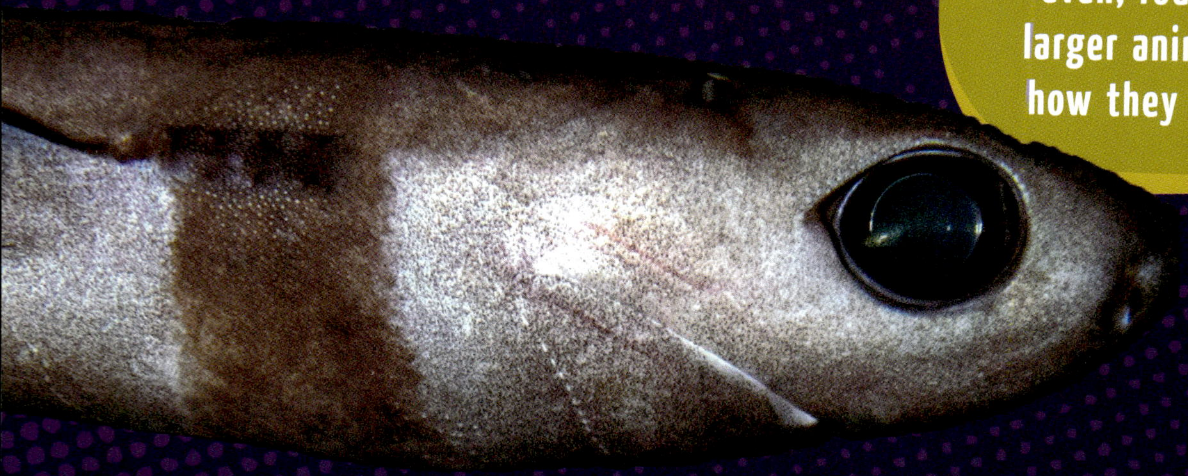

LOTS OF MARINE CREATURES ARE BIOLUMINESCENT, INCLUDING THE APTLY NAMED LANTERNFISH. HERE ARE A FEW OTHERS:

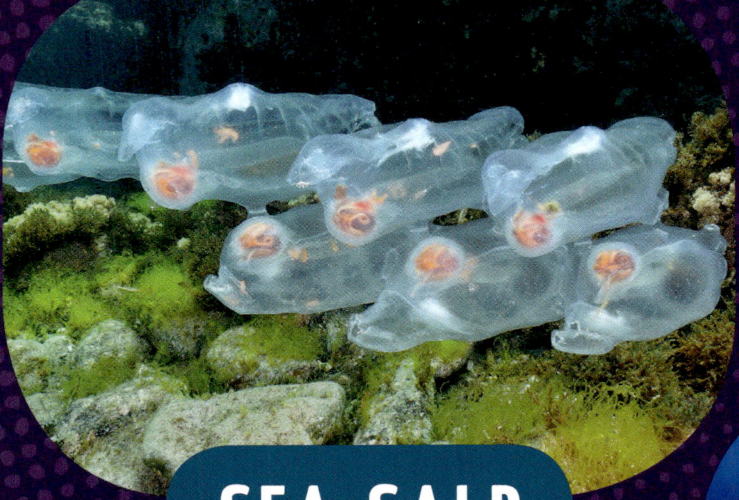

SEA SALP

JELLYFISH

HAWAIIAN BOBTAIL SQUID

WHAT TEETH!

DID YOU KNOW?

Sharks can have thousands of teeth at one time.

A shark needs its teeth! That's why shark teeth are constantly growing. They come in rows, and over time the smaller teeth in back move forward to replace those lost in front. Most sharks average between 5 and 15 rows of teeth.

Sharks are born with their teeth! And they might even need them before that— some kinds of sharks try to eat one another before they're born.

Unlike sharks, dolphins only get one set of teeth. The dolphins use them to capture their food—smaller fish, squid, and crabs—and then they swallow it whole. A bottlenose dolphin has between 72 and 104 teeth.

DID YOU KNOW?

Bottlenose dolphins have many teeth—but they don't use them for chewing.

MAGICIANS OF THE SEA

DID YOU KNOW?

Octopuses and their relatives are masters of disguise.

Octopuses and other **cephalopods** can change their skin **texture** and color, and they can match their surroundings really well. They also sometimes do the opposite and use bright colors to surprise predators. When this happens, they also may spray black ink. That allows them to swim away.

Octopus arms have suction cups! The octopus uses these to taste, grip, and smell.

If an octopus loses one of its eight arms to an attack or injury, it can regrow it! It takes a while—about 130 days or so—but in time it will be as good as new. Scientists are studying this ability to see if it can be used to help people who have lost limbs.

DID YOU KNOW?

Octopuses can regrow an arm after losing it!

DID YOU KNOW?

Mantis shrimp have split eyes.

Even though the mantis shrimp has only two eyes, they are split so that the marine animal can look above and below the waterline at the same time. It can also move each eye **independent** of the other. The shrimp have some of the best vision of any animal on the planet!

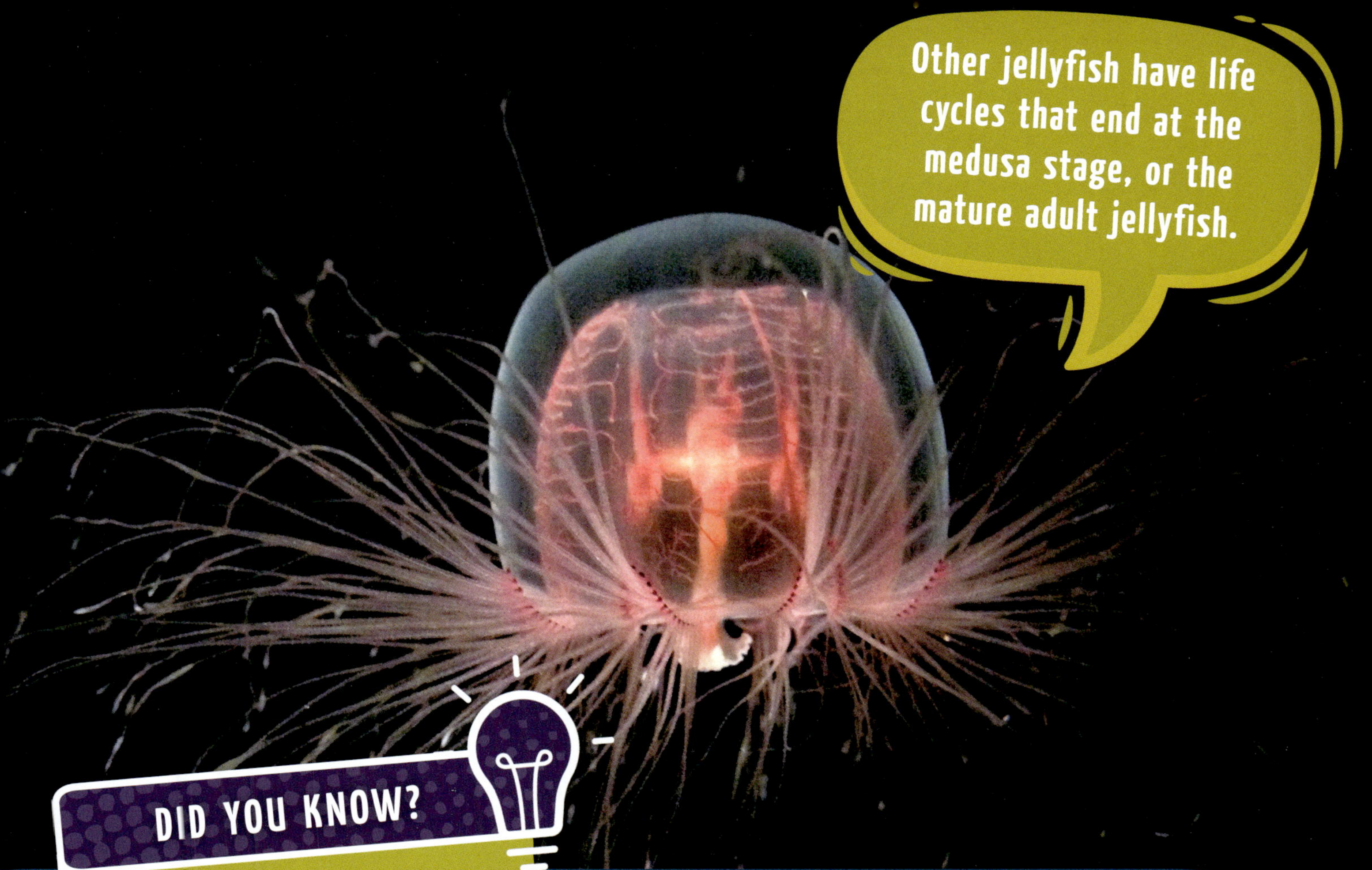

Other jellyfish have life cycles that end at the medusa stage, or the mature adult jellyfish.

DID YOU KNOW?

As its name suggests, the immortal jellyfish can live forever—or at least a very long time.

Most creatures go through a life cycle that ends in death. But not the immortal jellyfish! When it lacks food or when it's injured, the jellyfish turns back into a polyp—the stage of the jellyfish's life cycle between larva and adult. From there, it grows into an adult again. It can do this over and over.

DID YOU KNOW?

One sea star arm can regrow a whole sea star!

Some sea stars can regrow their arms if they lose one. But they can do more than that! The arms can regrow bodies too! As long as an arm is still attached to part of the center body, it can grow a whole new sea star.

Most sea stars have five arms, but they can grow as many as 50!

DID YOU KNOW?

Sea stars push their stomach out of their mouth to eat!

Sea stars eat mollusks, such as clams and oysters. They use suction cups on their arms to rip open mollusk shells. Then, their stomach comes out of their mouth to eat! When it's done with dinner, the stomach goes back inside the sea star.

MARINE FAMILIES

DID YOU KNOW?

Mother sea turtles remember the beach where they were born—and that's where they return to lay their eggs.

Scientists still aren't sure how mother sea turtles do it! When animals return to their birthplace to have babies, it's called natal homing. Mother sea turtles dig holes in the sand and lay their eggs. About two months later, the babies hatch, dig themselves out, and make their way to the sea.

It can take several days for baby sea turtles to dig out of their nests and make it to the sea.

DID YOU KNOW?

Baby sea turtles become male or female depending on the temperature around their eggs.

Temperatures inside the nest make a big difference! If temperatures are between 83°F and 85°F (28°C and 29°C), the sea turtle babies will be a mix of males and females. If it's colder than this range, they will all be males. If its higher than this range, they will all be females.

DID YOU KNOW?

Seahorses "date" one another before and after having babies.

Throughout multiple mating seasons, seahorses often stick with one partner. That way, they're more likely to produce the most offspring. Each day, male and female seahorses "dance" with one another.

When males give birth, it can be to as few as five or as many as 1,000 young seahorses.

DID YOU KNOW?

Male seahorses are the ones who give birth!

Female seahorses place their eggs into the male's pouch, which is called a brood pouch. This is where they are **fertilized**. It's also where they stay until the babies hatch. Because of this adaptation, female seahorses can begin producing more eggs right away.

DID YOU KNOW?

Sea hares (a sort of sea slug) can be both mothers and fathers at the same time.

All sea hares have both male and female **reproductive** parts. During mating season, they form large mating chains where many sea hares gather together. Afterward, each sea hare can lay millions eggs!

DID YOU KNOW?

Marine flatworms duel to decide who will carry the fertilized eggs.

Some marine flatworms are known as "Persian carpet flatworms" because of their colors and patterns.

Marine flatworms also have both male and female reproductive parts. When they meet, they fight! Neither one wants to carry the offspring, since it takes a lot of energy and effort. The loser is often injured in addition to having to carry the eggs.

SEA WONDERS

DID YOU KNOW?

The leafy sea dragon is so good at hiding that no other animals hunt for it!

These marine animals' bodies look just like the seaweed or kelp in which they live. They have appendages that look like leaves that sprout from their body.

Leafy sea dragons hide so well that they don't have any predators!

MANY SEA CREATURES USE CAMOUFLAGE AS A MEANS TO HIDE!

DECORATOR CRAB

SEVERAL SPECIES OF CRABS THAT STICK OBJECTS ON THEIR BODIES

STARGAZER FISH

BURY THEMSELVES IN SAND, LEAVING ONLY EYES AND MOUTH UNCOVERED

CUTTLEFISH

CHANGE THEIR COLOR, SHAPE, OR TEXTURE

UNDER THE SEA

An amazing variety of marine animals live under the water's surface. These creatures come in all different shapes and sizes, and many of them have developed special adaptations to survive their **habitat**!

Scientists continue to study the amazing life in Earth's oceans. They believe there's a lot more to learn, and there are many creatures that haven't even been discovered yet. To find out more about the ocean, visit your nearest aquarium, zoo, or natural history museum!

The ocean is huge! Scientists say 80 percent of it has yet to be explored.

GLOSSARY

appendage: Something (as a leg) attached to a larger or more important thing.

camouflage: Colors or shapes on animals that allow them to blend in with their surroundings.

cephalopod: A class of marine creatures including squids, cuttlefishes, and octopuses.

environment: The conditions that surround a living thing and affect the way it lives.

fertilize: To make an egg able to grow and develop.

habitat: The natural home for plants, animals, and other living things.

independent: Not controlled by something else.

nutrient: Something taken in by a plant or animal that helps it grow and stay healthy.

reproductive: Related to or involved in the creation of babies, young animals, or new plants.

substance: A material, or something from which something else can be made.

texture: The way something feels when you touch it.

threatened: Likely to be harmed.

toxic: Containing poison.

FOR MORE INFORMATION

BOOKS

Chapman, Blake. *Ocean Animals: The Weirdest, Smartest and Sneakiest Sea Creatures.* Clayton South, Australia: CSIRO Publishing, 2020.

Colosi, Rosie. *Alien Ocean Animals.* Washington, D.C.: National Geographic Kids, 2020.

WEBSITES

American Museum of Natural History: Ology
www.amnh.org/explore/ology/marine-biology
Want to learn all about marine biology? This website is jam-packed with videos, crafts, and games to help you learn all about life in the ocean.

National Geographic Kids: Ocean Habitat
kids.nationalgeographic.com/nature/habitats/article/ocean
Find out why the Earth's oceans are so important! Also discover some of the amazing creatures that live underwater.

Publisher's note to educators and parents: Our editors have carefully reviewed these websites to ensure that they are suitable for students. Many websites change frequently, however, and we cannot guarantee that a site's future contents will continue to meet our high standards of quality and educational value. Be advised that students should be closely supervised whenever they access the internet.

INDEX